Technology Inspired by ANIMALS

INVENTIONS INSPIRED BY REPTILES AND INVERTEBRATES

Tessa Miller and
Madeline Nixon

AV2
www.av2books.com

AV2

Step 1
Go to **www.av2books.com**

Step 2
Enter this unique code
JKFGQRJPB

Step 3
Explore your interactive eBook!

AV2 is optimized for use on any device

Your interactive eBook comes with...

Contents
Browse a live contents page to easily navigate through resources

Audio
Listen to sections of the book read aloud

Videos
Watch informative video clips

Weblinks
Gain additional information for research

Try This!
Complete activities and hands-on experiments

Key Words
Study vocabulary, and complete a matching word activity

Quizzes
Test your knowledge

Slideshows
View images and captions

... and much, much more!

View new titles and product videos at **www.av2books.com**

Technology Inspired by ANIMALS

INVENTIONS INSPIRED BY REPTILES AND INVERTEBRATES

Contents

Introduction to Biomimicry

Earth is crawling with creepy and slimy creatures. Scientists think there are more than 8 million different **species** of animals. Eighty percent of those are insects. This means Earth is covered with bugs. They come in all shapes and sizes. Worms can also be found everywhere. They crawl on or under the ground. They sometimes live inside other animals, including humans. There are also about 9,500 species of reptiles. These include snakes, lizards, turtles, alligators, and crocodiles.

That is a lot of creepy and crawly creatures. They have attracted the attention of many scientists. By studying worms, insects, and reptiles, scientists can come up with new **technology**. Scientists made a robot that moves just like a snake. They built a shopping mall modeled after a termite mound. Scientists and **engineers** are copying nature to improve lives. This is called biomimicry. "Bio" means life. "Mimicry" is when you copy something else. Many different technologies have been inspired by the creepiest and crawliest critters on Earth.

Young spiny-headed worms can live inside insects. When the insects are eaten by fish, the worms then move to live inside the fish.

Spiny-Headed Worms and Skin Grafts

Parasitic worms might not seem very helpful. For most people, just thinking about them makes their skin crawl. However, doctors have been inspired by these worms. The worms are helping to make medicine better.

Sometimes doctors see patients with very deep wounds. These wounds need to be sewn back together. Often, doctors sew wounds closed with stitches, or sutures. Sutures use a special type of sewing needle and thread. Sometimes the thread is not strong enough. Doctors then have to staple wounds closed, but a stapled wound can open up again.

Injuries, rather than illnesses, are the number-one reason kids visit the emergency room.

Doctors needed to find a better way to treat wounds where sutures and staples did not work. They looked to nature and found the spiny-headed worm. This worm latches onto its **host**. It can then stay strongly attached. By mimicking the worm's abilities, doctors are developing a new type of medical patch. The patch would latch on just like a spiny-headed worm. It could hold wounds together more securely. It could also help keep wounds cleaner. This would speed up recovery time.

Lessons from Nature

Worms can be found in fish, amphibians, reptiles, birds, and mammals.

The spiny-headed worm uses fish as a host. It lives in the fish's intestines. This worm has a special head. It is shaped like a needle. First, the worm sticks its head into the intestine. The head then swells up. This keeps the worm stuck in place.

Doctors copied this. They are making a patch that is covered with hundreds of tiny hooks. Each hook acts like a worm's spiny head. The patch is gently pressed into the skin around the wound. The hooks then swell up. This holds the wound closed.

A team of researchers at Brigham and Women's Hospital are working on skin graft technology inspired by spiny-headed worms.

Doctors have not tried the patch on patients yet, but they hope the tiny hooks are too small to hurt. They want to use the patch on burn patients. These patients often require skin **grafts**. The patch would hold the grafts more securely.

Doctors also hope to use the patch on wounds inside the body. The patch would work better than sutures. It would stick to these surfaces, just like a parasitic worm.

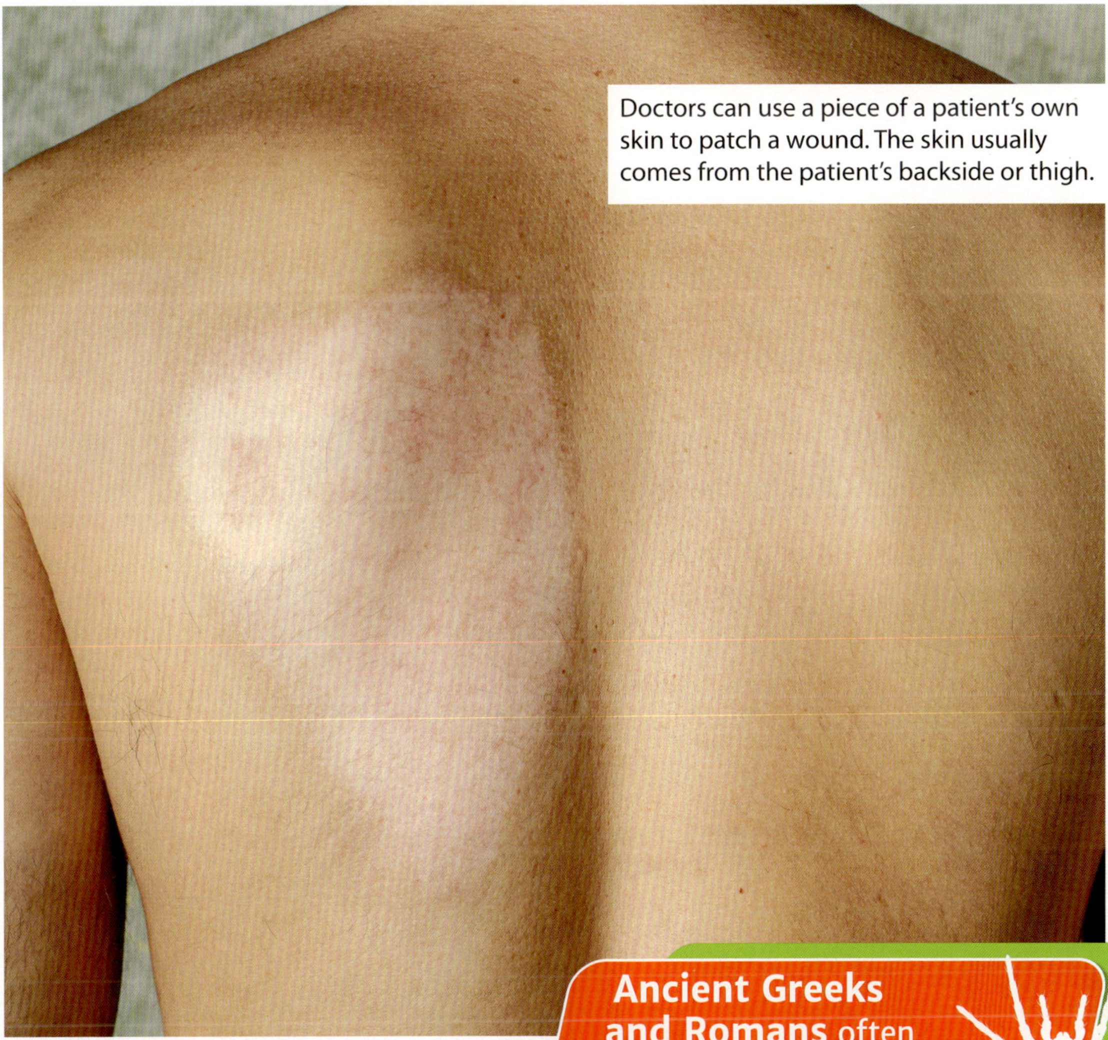

Doctors can use a piece of a patient's own skin to patch a wound. The skin usually comes from the patient's backside or thigh.

Ancient Greeks and Romans often used **spider silk** to cover wounds.

Tech in Action

Parasitic worms can make people sick. But doctors are using them as inspiration for improving medicine.

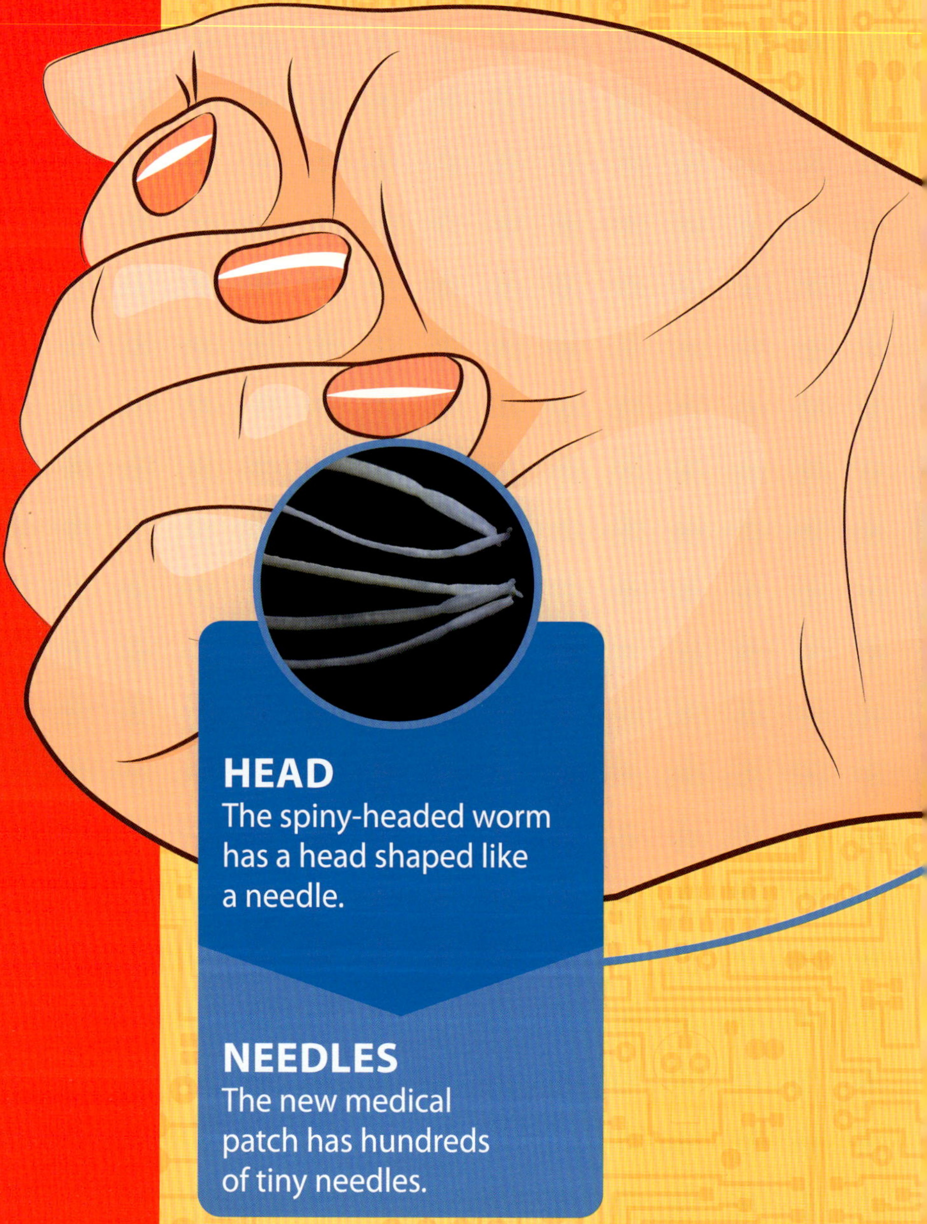

SWELLING
The head swells up after poking a hole in the intestines of a fish.

STICKING
The needles swell up when they get wet.

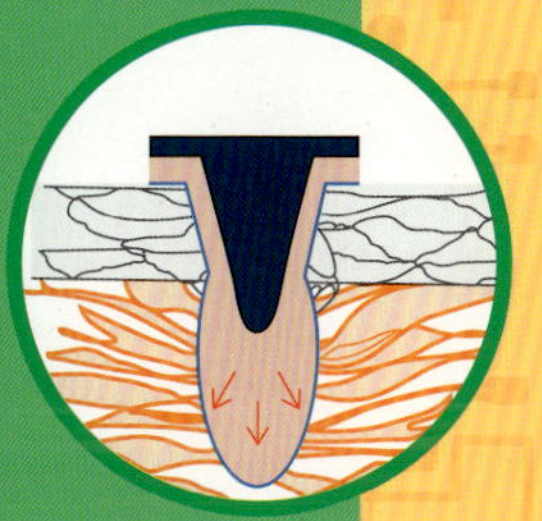

IN PLACE
The swelling keeps the worm in place until it decides to move.

SECURE
The patch stays in place even on wet surfaces.

Termites use soil and their own saliva and droppings to build their mounds.

Termite Mounds and Buildings

Have you ever walked upstairs and found that the temperature is hotter than downstairs? That is because hot air rises. When air heats up, it becomes less **dense**. The hot air then floats on top of the cold air. This can be a problem in tall buildings. The hot air rises to the top floors. In very hot places, cooling down the top floors can be tough.

Zimbabwe's Eastgate Centre does not have air conditioners. It stays cool thanks to biomimicry.

Today, air conditioners are usually used to cool down tall buildings. Sometimes air conditioners run all day and night. This uses a lot of electricity and can be very expensive. Mick Pearce, an architect in Zimbabwe, wanted to solve this problem. Architects are people who **design** buildings.

Pearce looked to nature for a solution. He found termite mounds. These mounds stay cool inside during the hottest part of the day. They also stay warm during the coldest part of the night.

A termite mound is very complex. It includes many tunnels and rooms for food storage.

Lessons from Nature

Termites keep their mounds cool with a unique **system**. In the middle of the mound, there is a chimney. Hundreds of tiny tunnels are attached to it. The tunnels lead to the outside. During the day, the tunnels fill with hot air. This hot air pushes the cool air down. The cool air goes into the living area of the mound. During the day, the air in the living area slowly warms up. At night, cooler air enters the tunnels. The warmer air in the living area rises. It pushes the cool air back down. This air **current** keeps the living area at a livable temperature.

Termite mounds can grow as high as 17 feet (5 meters).

Pearce designed Zimbabwe's largest shopping mall, the Eastgate Centre. He copied the termite's system. The roof has vents that act like chimneys. The walls have tubes. They run to the outside of the building. This system creates air currents like a termite mound. It keeps the building cool. The building does not need to run air conditioners all the time. This saves a lot of money.

The Eastgate Centre is home to many shops and business offices. It covers half a city block.

In the United States, it costs about $100 to $300 a month to cool down a house using air conditioners.

In one day, temperatures in Zimbabwe can swing from **35° Fahrenheit to 104°F (2° Celsius to 40°C).** That is a difference of **70°F (38°C).**

Tech in Action

How do you keep a large building cool without using a large amount of energy? Biomimicry of a termite mound is one way.

CHIMNEY

Termite mounds have a central chimney.

VENTS

Vents in the Eastgate Centre's ceiling mimic a termite mound's. They let hot air move upward.

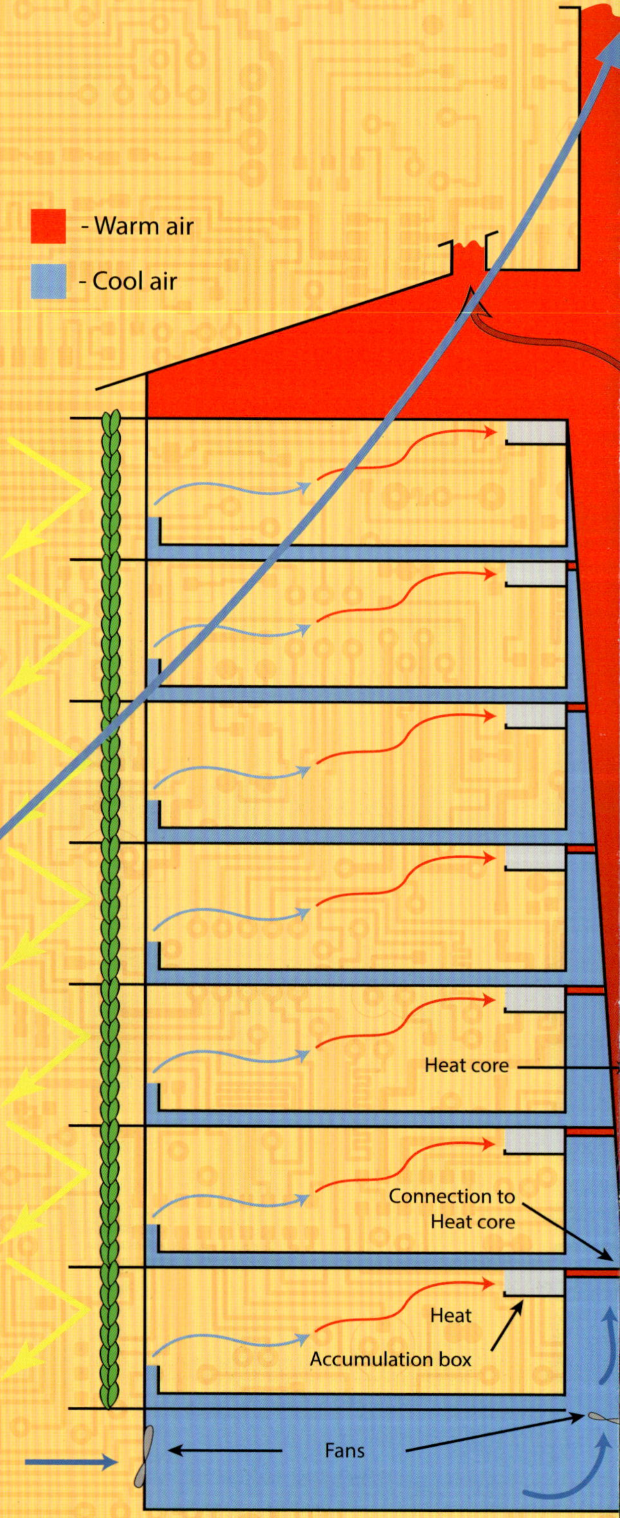

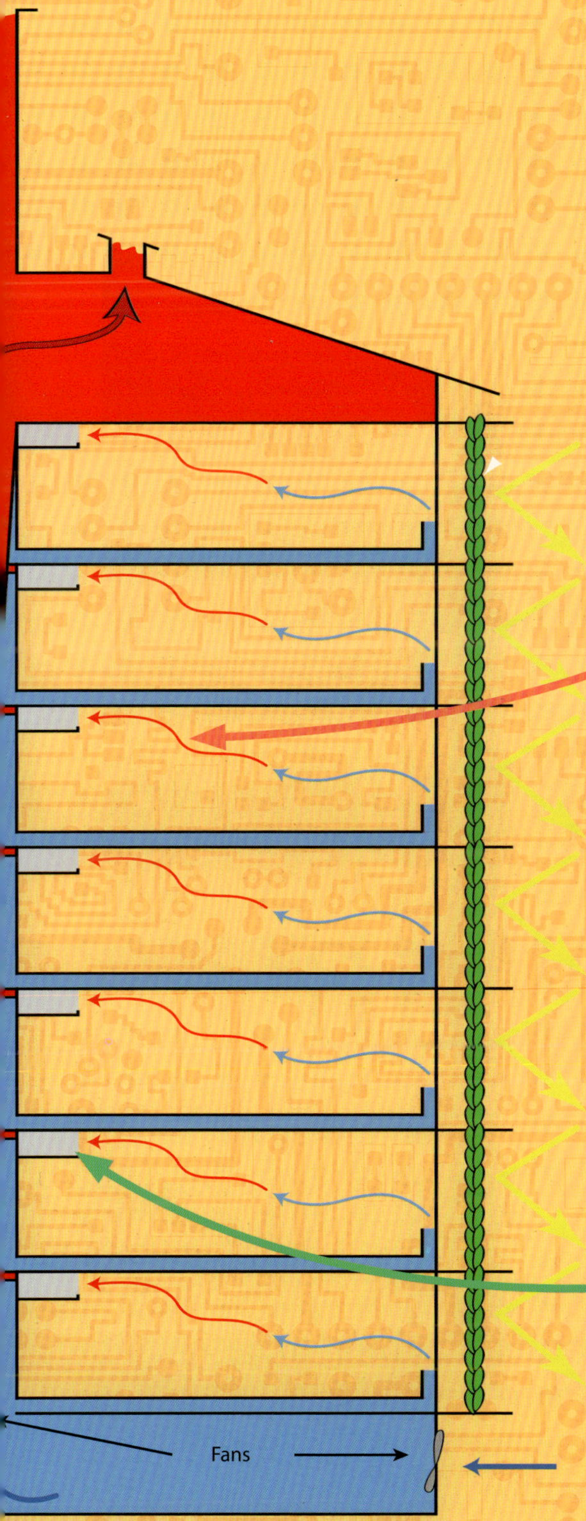

AIR MOVEMENT

Because hot air rises and cold air sinks, the air inside the termite mound is constantly moving.

AIR FLOW

The Eastgate Centre has constant air movement. This keeps it cool.

TUNNELS

The tunnels attached to the central chimney fill with hot or cold air.

TUBES

There are hundreds of tubes in the walls of the mall. They let hot and cold air move.

A spider makes silk for spinning webs, catching bugs, and laying eggs. The silk is made of proteins.

Spider Webs and Bulletproof Vests

In recent years, there have been wars in some very hot places. Many soldiers fought in the Iraq War (2003–2011). They were faced with temperature highs of 110°F (43°C) in the city of Baghdad.

Soldiers who work in such heat want their clothing to be as light as possible. Instead, they have to wear very thick and heavy uniforms. The uniforms must cover all parts of their bodies.

Bulletproof vests are usually made of Kevlar. Kevlar is a type of very strong plastic.

Soldiers have to wear heavy body armor as well. Body armor protects against bullets. It keeps soldiers safe. It can also make working in hot places very difficult.

Fortunately, there are scientists working for the U.S. military. They are looking at ways to reduce the weight of a soldier's body armor. They are working with one of the strongest **fibers** in the natural world—spider silk.

Spider silk is liquid until it hits the air. Then, it hardens into a stretchy, strong fiber.

Lessons from Nature

Spiders make their webs from spider silk. The silk is very stretchy. It can be pulled farther before breaking than any other natural fiber. The silk is very strong when woven. These thin fibers are also extremely lightweight.

Scientists in the Netherlands have a **lab** crawling with spiders. They created a **material** that mimics spider silk. It looks like human skin. It can stop bullets. In a lab in Michigan, researchers use spiders and silkworms to make very strong silk. Using that silk, they invented a new fabric called Dragon Silk. This fabric could be used to make better bulletproof vests. Dragon Silk is stronger than the vests police officers and soldiers wear today. Making it is also better for the **environment**. No chemicals are needed and no pollution is created.

A company in Japan is using microorganisms, tiny living things, to grow silk like a spider's. Now the company can make silk very quickly. They are able to use it to make car parts that are strong but flexible. These parts can help keep people from getting hurt in accidents.

Many research groups and companies around the world are competing to make products from spider silk.

Silkworms build cocoons out of a **single strand of silk**. These strands can be **3,000 feet (900 m) long**.

Tech in Action

Spider silk is a unique natural fiber. Scientists are coming up with new ways to use it to help humans.

STRONG
Spider silk is stronger than steel when woven together.

TOUGH
Dragon Silk is so strong it can stop bullets.

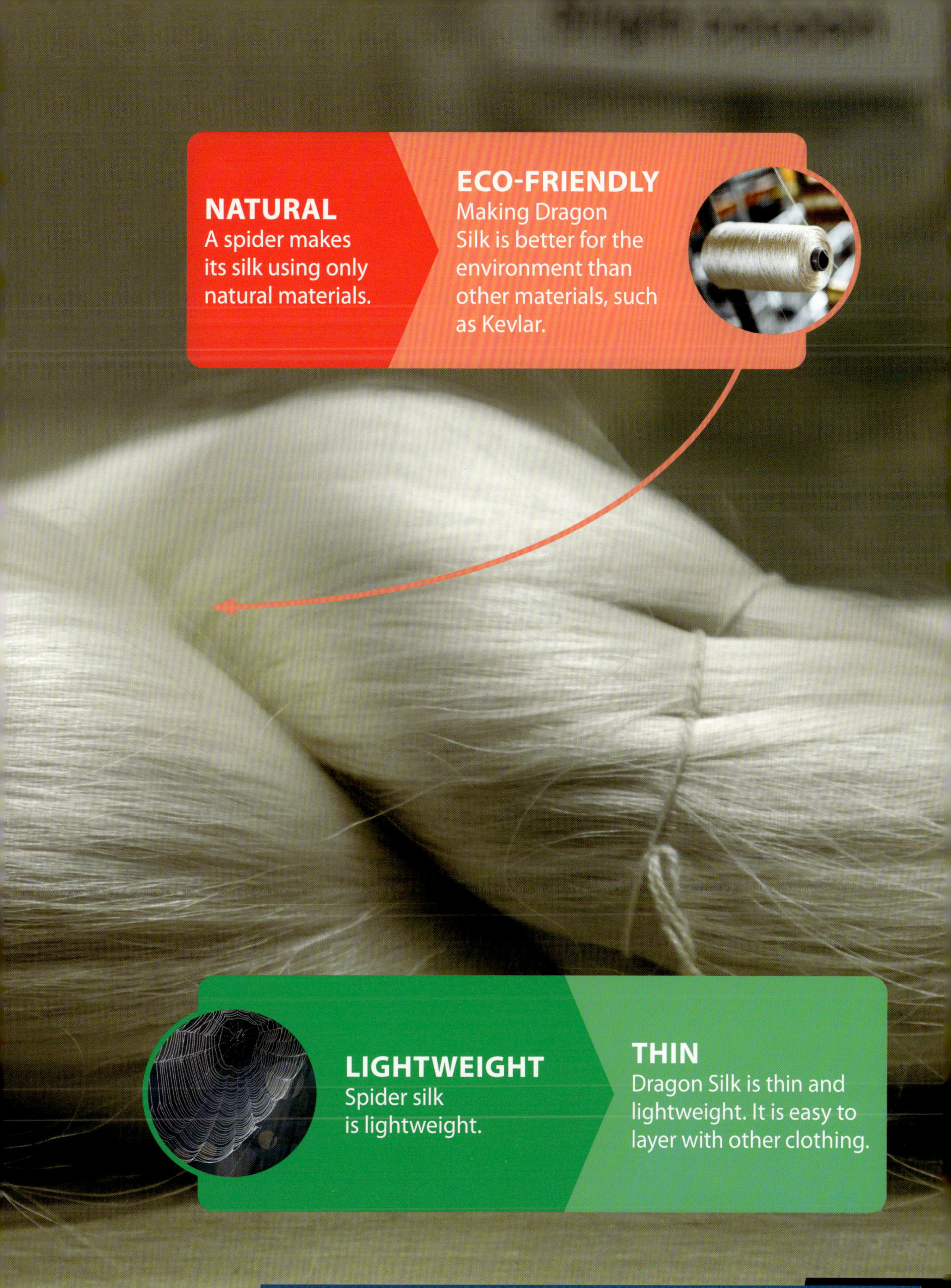

NATURAL
A spider makes its silk using only natural materials.

ECO-FRIENDLY
Making Dragon Silk is better for the environment than other materials, such as Kevlar.

LIGHTWEIGHT
Spider silk is lightweight.

THIN
Dragon Silk is thin and lightweight. It is easy to layer with other clothing.

Geckos are a type of lizard. There are almost 1,000 different species of geckos.

Geckos and Adhesive

Geckos are some of the best climbers in the world. These tiny lizards can climb almost anything. They can crawl across trees and ceilings. They can creep along flat and slippery surfaces. What would it be like if a human could climb like this? What if a person could climb a wall using only their hands and feet?

New high-tech fabrics and materials are called "smart clothes" or "wearable technology."

A U.S. military research group wanted to answer these questions. They started a research **program** and hired **biologists** to study geckos. From their studies, they have made a new material. They call it Geckskin. This material could be worn by soldiers during missions. It would allow a soldier to climb just like a gecko. They would not need complicated, heavy climbing gear. They would not have to pack rope and a harness. Soldiers could just slip on the Geckskin and climb.

Geckos can latch and unlatch their feet from surfaces very quickly. This lets them run away from danger.

Lessons from Nature

Geckos' feet are covered in millions of **microscopic** hairs. These hairs act like little hooks. As the gecko walks, **tendons** that go from the toes to the hairs tense up. The gecko is now latched on. When the gecko wants to move, the tendons relax. The hairs are released, and the gecko moves along.

Geckskin is built to work like this. It has a soft layer of **microfibers**. It mimics a gecko's toes. That layer is covered with a stiff fabric. The fabric acts like the gecko's hairs. It is woven in a special way. This lets it work like the gecko's tendons. A small patch of Geckskin can hold up to 700 pounds (320 kilograms).

The military wants soldiers to use Geckskin to climb buildings in cities. They would not have to carry ropes and ladders. With Geckskin, they could move much more quickly and safely.

Scientists plan to use Geckskin for more than climbing up walls. They want to make robots that climb around like a gecko. These robots would climb around in space, on the outside of a spaceship.

If a gecko used **every micro-hair** on its feet at the same time, it could support **280 pounds (127 kg)** of weight.

Tech in Action

A gecko can crawl and climb using only its feet. By wearing Geckskin, people can do the same thing.

TINY HAIRS

Geckos have thousands of tiny hairs on their feet. These act like hooks.

MICROFIBERS

Thousands of microfibers help Geckskin cling to a surface.

TENDONS
Each hair has a tendon. It forces the hair to cling to a surface.

FABRIC
A layer of stiff fabric holds each microfiber in place.

TRACTION
Geckos can climb vertical surfaces, such as walls. They can even climb on ceilings.

GRIP
With Geckskin, a person can climb straight up or hang upside down.

There are more than 3,000 different species of snake. Since none of them have legs or feet, they use their muscles and scales to move.

Snakes and Robots

A building has collapsed after a powerful earthquake. Rescuers have been sent in to search for people. Survivors are crying out for help. But the rescuers cannot reach them. The way is blocked by **debris**.

Luckily, scientists at a lab in Pennsylvania have designed a robot that can help. It is shaped like a snake. This robot can slither into small places. It can also climb. It can move faster than even the fastest snake.

Labs around the world are making new robots inspired by snakes.

Using the robot, searchers can stay safe while the robot winds its way through danger. It can be **programmed** to explore areas on its own. People can also control it with a remote.

The researchers who invented these robots do not call their work biomimicry. They are inspired by nature, but they are not copying it. They are taking something in nature and making it even better. They call this "bioinspiration."

Lessons from Nature

The lab calls its robots **modular** snake robots. They tested the first modular snake robot in many different places, from rainforests to frozen lakes. It could successfully move over almost any type of land. But it did not work in deep sand. It could not gain traction when the sand moved under its weight. So scientists had to find an animal that could walk on top of sand.

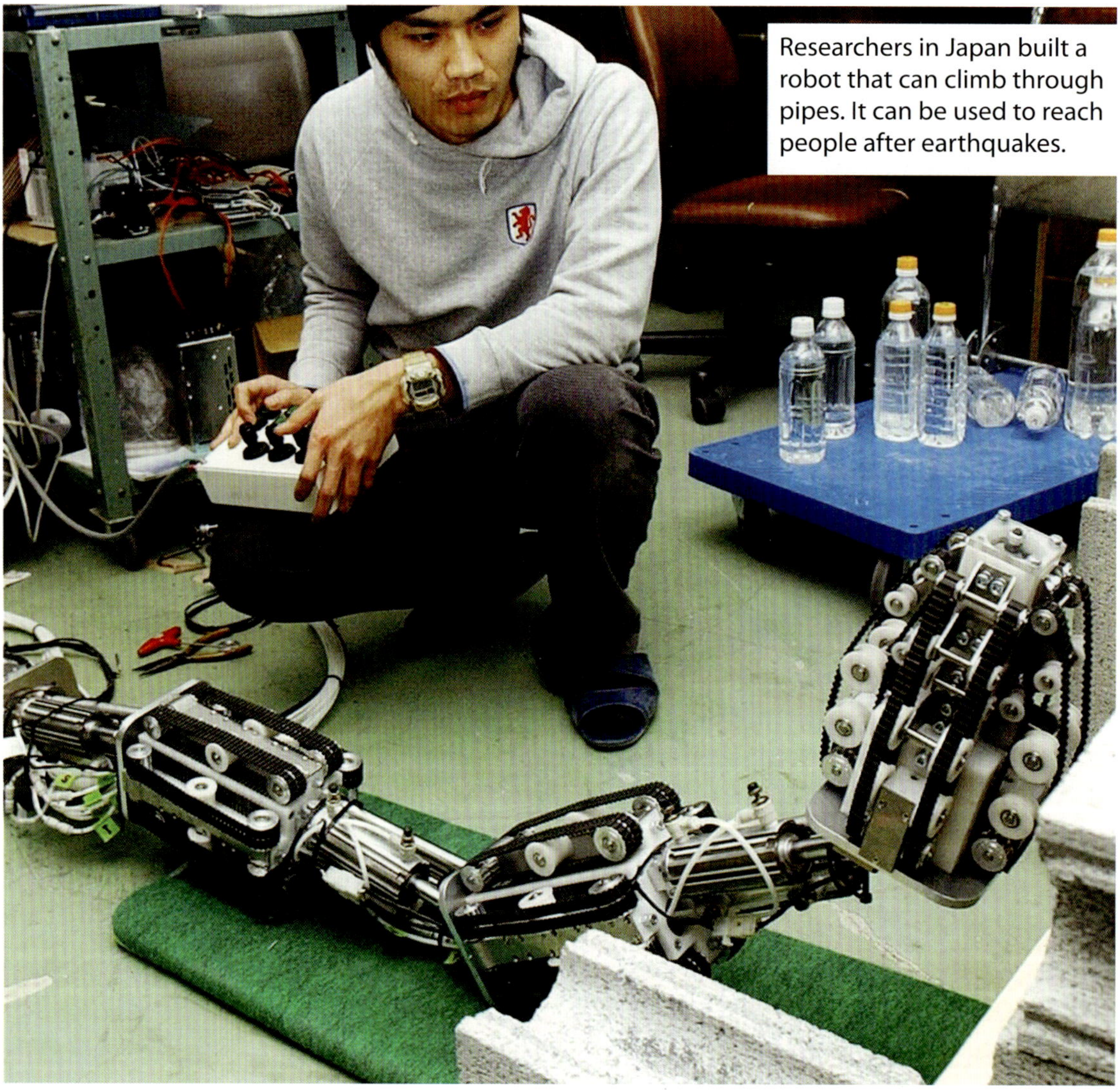

Researchers in Japan built a robot that can climb through pipes. It can be used to reach people after earthquakes.

Researchers looked to sidewinder snakes in Arizona for help. These snakes have a unique way of slithering. They squish their bodies as flat as they can. Then they twist in a zigzag pattern. The motion lets the snakes travel across the sand, even when the sand is moving. It also lets them easily climb sand dunes.

This was a **breakthrough** for the scientists. They created a modular snake robot that mimics the sidewinder's twist. This robot can now go anywhere. It can be used for search and rescue. The robot could save many lives during a disaster.

Sidewinders are a type of rattlesnake. Their bites are venomous.

A snake robot is **built in sections**. Each section **carries special items**. People can **add sections to the** robot if it needs to do more than one thing on a mission.

Tech in Action

To build a search-and-rescue robot, engineers mimicked the body and movement of a snake.

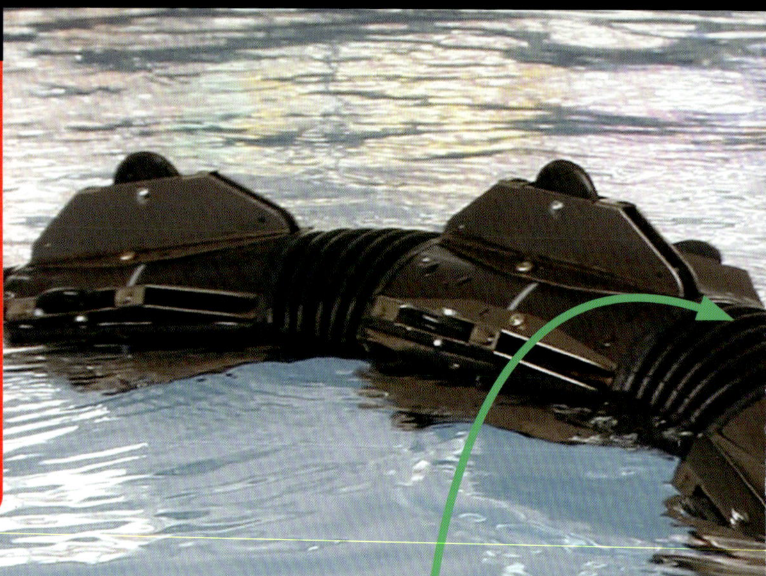

SURFACES
Sidewinder snakes can zigzag up steep sand dunes.

AREAS
Scientists can program the robots to swim, climb, and zigzag through any area.

FORM
Snakes are skinny and very flexible.

SHAPE
Scientists made the modular robot in the shape of a snake.

MANEUVERABILITY
Snakes move forward and backward, swim, and climb without feet or hands.

MOBILITY
While some use tracks, snakelike robots move without wheels or wings.

Ants are excellent problem solvers. Engineers, architects, math experts, and computer scientists study them to solve human problems.

Ants and Antivirus Software

Ants are not very big. The common black ant is only about 0.2 inches (0.5 centimeters) long. Ants have to work together to survive. They live together in **colonies** or nests. There are hundreds, even thousands, of ants per colony. Each ant has a job. By working together and doing their jobs, ants can be very powerful. Sometimes, other animals attack an ant colony. The colony fights back.

Computers work like an ant colony. There are thousands of pieces in a computer. This includes hardware, such as the screen and keyboard. It also includes **software**. Each piece has its own job. When the pieces are working together, computers can get a lot done.

When something attacks a computer, such as a virus, it cannot fight back. Computer attacks are becoming a big problem around the world. Researchers work hard to make software that fights back. One new software mimics how ants protect their colony.

Computer viruses are malicious. This means they are made to cause harm, such as stealing information.

Lessons from Nature

People already use software to protect their computers. The software **scans** for viruses. The problem is, if it scans all the time, it slows down the computer. It can also scan once a night or once a week. This leaves the computer open to attack during the time the program is not scanning. Researchers from North Carolina and Washington state made a new antivirus software. It uses digital "ants" to protect the computer.

In an ant colony, there are scouts. They go off in search of food. If they find something, they go back to the colony and signal to the other ants. Then, a large group goes to collect the food. Ant colonies also have soldier ants. They attack things that might hurt the colony. Like the ant colony, the antivirus software first sends out scouts. These digital scouts wander the computer system. If they find a virus, they send signals to other digital ants. Those ants then attack the virus.

Soldier ants attack to protect their colony. They might also attack to get food.

After an attack on an ant colony, builder ants come to fix any damage. The new software does this, too.

This new software scans constantly. It only uses a few digital ants at a time. It does not slow a computer down. This means it can send lots of ants in quickly to attack when they are needed.

Large companies often hire computer scientists. They work to protect computer systems from viruses and other attacks.

One of the **largest known ant colonies** is in Europe. It stretches for **3,700 miles (6,000 kilometers) underground.**

Tech in Action

Ants are small, but they are powerful when in a large group. A new antivirus software mimics how ants protect their colony.

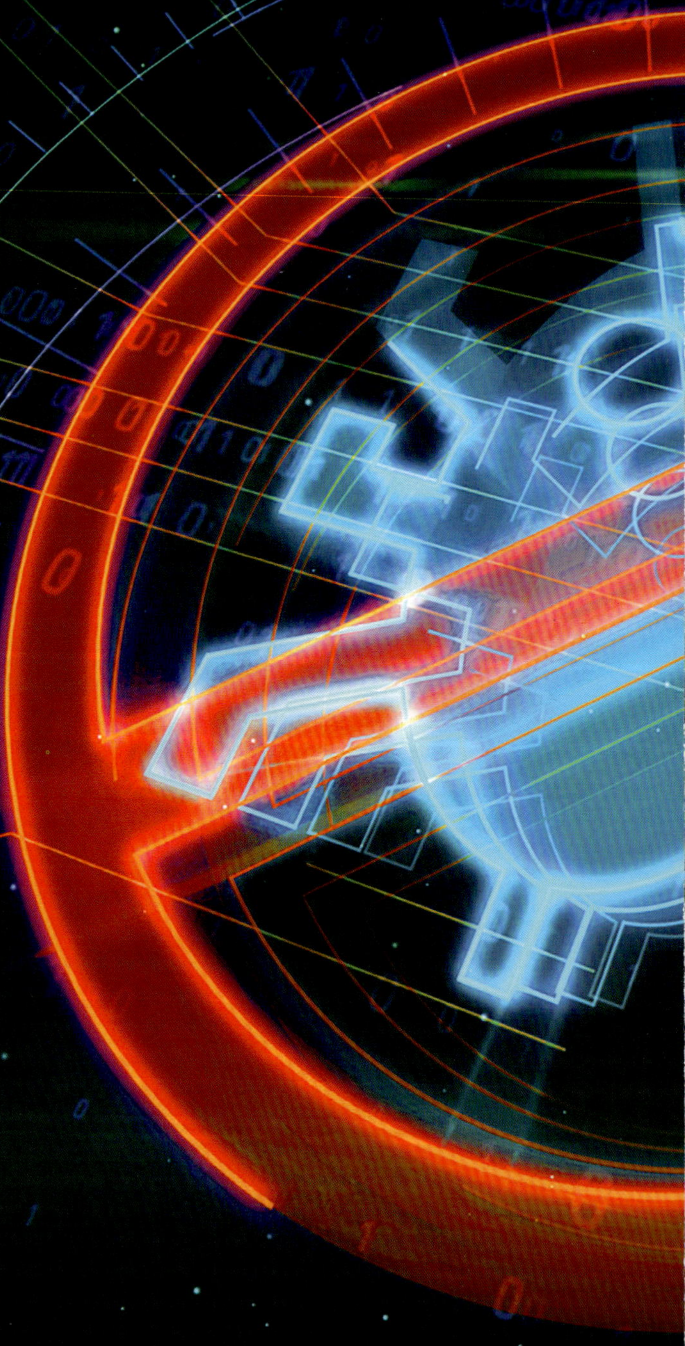

PATROL
Scout ants constantly patrol the colony.

MONITORING
Antivirus software "ants" monitor the computer's systems.

SWARM
When there is a problem, soldier ants swarm together to help fix it.

ATTACK
If a virus is detected, the software sends many more "ants" to attack it.

REPAIR
Builder ants repair any damage done to the colony by the attack.

FIX
After the attack, the software repairs the computer.

A snake flicks its tongue to taste chemicals in the air. It uses the information to build a map of its environment.

Conclusion

Even the creepiest animals can improve technology. The study of worms, insects, and reptiles has made people's lives better. It has led to many new inventions, including medical patches that latch wounds together. These patches could help people heal faster after injury. Geckskin could allow humans to scale walls like a gecko. Bulletproof vests of the future could be made of spider silk. These technologies would not be possible if scientists had not used biomimicry.

Scientists continue to build snake robots. New robots can respond to chemicals in the environment.

New technologies are exciting and can do a lot of good. Unfortunately, they are also **risky**. Snakelike robots could be used by criminals. Geckskin suits could also be used for spying or to help break into buildings.

Scientists continue to change the world by studying creepy and crawly creatures. Who knows what new inventions might creep up next?

Timeline

Biomimicry has helped to create numerous types of technology. Reptiles and insects have given inventors inspiration for inventions.

1942

Igor Sikorsky studies dragonfly wings and flight. Dragonflies are capable of hovering, flying backwards, and making many other difficult maneuvers. Sikorsky applies what he learns and creates the first single-rotor helicopter that is stable and fully controllable.

2009

Researchers study gecko eyes and find that they have a high number of cones in their eyes. This means they can see color better and at night. The discovery has been applied to contact lenses and cameras for higher definition.

2011

A Japanese scientist streamlines needles by studying mosquitoes. His goal is to create a needle that will be less painful and less noticeable, just like a mosquito bite.

2017

While studying a dark black butterfly, Dr. Hendrik Holscher discovers that its color is perfect for heat absorption and management. Applying this discovery to solar panels will allow for more green energy.

2018

Kyoo-Chul Park and his team of researchers create a liquid-collecting permeable structure, or water tent. This invention is inspired by the darkling beetle, which lives in the Namib Desert. The darkling beetle collects fog on its wing tips and turns it into water for itself.

2019

Chang-Jiang Chen researches fireflies so that LED lights can become more energy efficient. Fireflies have asymmetrical microstructures on their bodies. These help them spread their light instead of keeping it in one area.

Activity

Explore Animal Feet

How do mountain goats cling to rocks? How do penguins keep from slipping on ice? Try this activity to explore how animals' feet help them gain traction on different surfaces. You can test different shoes to figure out how they grip.

What You Need

- Shoes with different types of soles. These can be cleats, ballet slippers, tennis shoes, high heels, hiking boots, snow boots, flip flops, etc.
- Different surfaces to walk on. These can include carpet, a gym floor, linoleum, wood, pavement, gravel, grass, mud, etc.
- A pencil and paper

What to Do

1. Break up into equal teams. In your group, examine at least three pairs of shoes. Write down what surface you think they would be best for walking on.
2. Brainstorm animals from different environments. Think about what types of feet they have. Compare them to the shoes you examined. How are they similar? How are they different?
3. Next, conduct a walking experiment. Have a group member walk on each surface. Use each pair of shoes. Keep track of which shoes worked best on each surface, being careful not to slip and fall.
4. After all teams have tried this, compare notes.
5. Which shoes worked best on which surfaces? Was there a shoe that did not help you walk on any surface? Return to your list of animals. Which animal would walk best on these surfaces? Do their feet match the soles of the shoes?

Key Words

biologists: scientists who study living things, such as plants and animals

breakthrough: a sudden discovery that leads to an increase in knowledge

colonies: groups of people or animals that live together

current: a flow of air or water in one direction

debris: remains left after destruction

dense: made up of pieces that are packed tightly together

design: to make a plan by thinking about the purpose or use of something

engineers: persons who plan and build tools, machines, or structures

environment: the air, water, plants, animals, weather, and other things in an area

fibers: thin threads that are usually used to make something, such as fabrics

grafts: pieces of skin, muscle, or bone that are moved from one part of the body to another to help it heal

host: the animal or plant in which a parasite lives

lab: a laboratory; a place for doing scientific work

material: cloth or fabric, usually before it is made into something

microfibers: very fine fibers that are not natural but are made by humans

microscopic: very tiny; only able to be seen with a microscope

modular: made up of many pieces that can come together in different ways

parasitic: living inside of and getting food from another animal or plant

program: a plan or project set up to achieve a long-term goal

programmed: given a set of instructions to perform a certain task

risky: having a chance of something bad happening

scans: looks at or studies carefully in order to find something

software: computer programs

species: a group of plants or animals with similar features

system: a set of parts that work together

technology: tools and knowledge used to meet a need or solve a problem

tendons: stringy pieces of tissues connecting muscles, bones, and other parts of the body

Index

Get the best of both worlds.

AV2 bridges the gap between print and digital.

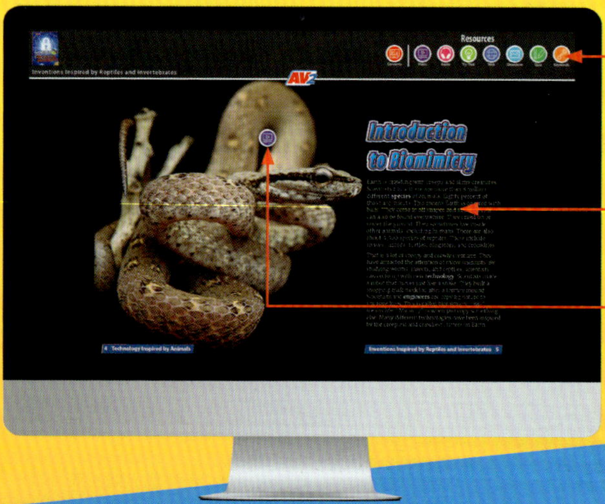

The expandable resources toolbar enables quick access to content including **videos**, **audio**, **activities**, **weblinks**, **slideshows**, **quizzes**, and **key words**.

Animated videos make static images come alive.

Resource icons on each page help readers to further **explore key concepts**.

Published by AV2
14 Penn Plaza, 9th Floor New York, NY 10122
Website: www.av2books.com

Library of Congress Cataloging-in-Publication Data

Names: Miller, Tessa, author. | Nixon, Madeline, author.
Title: Inventions inspired by reptiles and invertebrates / Tessa Miller and Madeline Nixon.
Description: New York, NY : AV2, 2021. | Series: Technology inspired by animals | Includes bibliographical references and index. | Audience: Ages 11-14 | Audience: Grades 4-6
Identifiers: LCCN 2019046689 (print) | LCCN 2019046690 (ebook) | ISBN 9781791118167 (library binding) | ISBN 9781791118174 (paperback) | ISBN 9781791118181 | ISBN 9781791118198
Subjects: LCSH: Biomimicry--Juvenile literature. | Inventions--Juvenile literature.
Classification: LCC T173.8 .M548 2021 (print) | LCC T173.8 (ebook) | DDC 600--dc23
LC record available at https://lccn.loc.gov/2019046689
LC ebook record available at https://lccn.loc.gov/2019046690

Printed in Guangzhou, China
1 2 3 4 5 6 7 8 9 0 24 23 22 21 20

052020
101519

Art Director: Terry Paulhus Project Coordinator: Priyanka Das

The publisher acknowledges Getty Images, iStock, Minden, Shutterstock, and Wikimedia as its primary image suppliers for this title.

First published by Full Tilt Press in 2019.